ELEMENTS OF THE PERIODIC TABLE

WHAT IS OXYGEN?

KATHLEEN A. KLATTE

Published in 2026 by The Rosen Publishing Group, Inc.
2544 Clinton Street, Buffalo, NY 14224

Portions of this work were originally authored by Mary Thomas and published as *Oxygen*. All new material this edition authored by Kathleen A. Klatte.

Designer: Rachel Rising
Editor: Kathleen A. Klatte

Library of Congress Cataloging-in-Publication Data
Names: Klatte, Kathleen A., author.
Title: What is oxygen? / Kathleen A. Klatte.
Description: Buffalo, New York : Rosen Publishing, 2025. | Series: Elements of the periodic table | Includes bibliographical references and index.
Identifiers: LCCN 2024054092 | ISBN 9781499478495 (library binding) | ISBN 9781499478488 (paperback) | ISBN 9781499478501 (ebook)
Subjects: LCSH: Oxygen--Juvenile literature. | Chemical elements--Juvenile literature.
Classification: LCC QD181.O1 K53 2025 | DDC 546/.721--dc23
LC record available at https://lccn.loc.gov/2024054092

Some of the images in this book illustrate individuals who are models. The depictions do not imply actual situations or events.

Manufactured in the United States of America

CPSIA Compliance Information: Batch #CSRYA26. For further information, contact Rosen Publishing at 1-800-237-9932.

CONTENTS

INTRODUCTION

Since ancient times, people have looked up at the night sky and wondered if there was life out there. Today we know that one of the key components of life as we know it is oxygen. Advanced scientific exploration has revealed that out of all the celestial bodies in our solar system, only Earth currently has enough oxygen in its atmosphere to support life. Earth's atmosphere is composed of about 21 percent oxygen. This, together with the planet's distance from the sun and a strong enough gravitational field to hold the atmosphere in place, makes our planet unique.

It was once thought that Mercury had a high percentage of oxygen in its atmosphere, but its proximity to the sun and weak gravitational field meant it couldn't form and hold a substantial enough atmosphere to support life. Subsequent space missions using more sophisticated equipment have revealed that there isn't as much oxygen in Mercury's atmosphere as previously thought.

However, decades of exploration have revealed that Mars might once have had enough oxygen to support life as we know it. NASA's *Curiosity* rover discovered rocks with a high content of manganese oxide. This is a mineral that was present on Earth when the earliest life forms began in the sea. Its presence on Mars suggests that there may once have been enough oxygen in the atmosphere for life to exist.

More incredible still, it may be possible to extract oxygen from the thin Martian atmosphere. NASA's *Perseverance* rover carried an instrument called the Mars Oxygen In-Situ Resource Utilization Experiment (MOXIE). MOXIE was used for a series of experiments in partnership with the Massachusetts Institute of Technology (MIT). MOXIE uses electricity to split carbon dioxide from Mars's atmosphere into carbon monoxide and oxygen molecules.

This technology is important because oxygen is necessary for humans to breathe and also to burn rocket fuel. For humans to exist on Mars and eventually return to Earth, they would either need to carry millions of tons of oxygen with them or manufacture it on Mars.

What makes oxygen so important? A closer look at this element reveals why it is so essential.

Mars has polar ice caps, just like Earth. Some of it is dry ice (frozen carbon dioxide), but some of it is water ice. Oxygen is a major component of water, so this is further evidence that Mars once supported a richer atmosphere.

CHAPTER 1

THE AIR WE BREATHE

Oxygen (O) has no taste, smell, or color, yet it is all around us—in the air we breathe and the water we drink. Oxygen is one of the reasons why Earth has an atmosphere that supports life as we know it. It's also a major component of the water that fills our seas. Without oxygen, the surface of the earth would be dry and empty like Mars or the moon.

Oxygen is an element—that is, a substance composed of only one kind of atom. Elements are the building blocks of life. They combine in different ways to form everything that lives on Earth. There are over 100 elements currently recognized by scientists. More are discovered as technology improves. Most of these elements have only been discovered in the last two centuries. The periodic table of elements, the system we use to organize them, is about 150 years old. It's currently believed that people have discovered all the naturally occurring elements. Other elements are man-made. They're the result of lab experiments, usually involving radiation. For a new element to be included on the periodic table, it must undergo a rigorous examination process.

Different scientists in different labs must be able to reliably recreate the same substance.

Oxygen easily forms compounds with other elements. Its most important compound is water. Water covers about 70 percent of the earth's surface. Our bodies are also about 60 percent water. We breathe in oxygen and exhale carbon dioxide. Plants take in carbon dioxide and release oxygen. This is one of the cycles that depends on oxygen. Another involves water evaporating, forming clouds, and then falling back to Earth.

The word "scientist" was first used in 1834. There were people we would consider modern scientists performing experiments to prove their theories by the 16th century and sometimes long before. However, until the 18th century, it was still common for natural philosophers to discuss their theories in the abstract without trying to prove them.

Credit for discovering oxygen is shared by scientists who conducted independent experiments in different countries. About 1772, Carl Wilhelm Scheele (1742–1786) of Sweden discovered oxygen, which he called "fire air," while experimenting with nitric acid. When he heated nitric acid, a gas was released that made nearby candles burn brighter. Around 1774 in England, a chemist named Joseph Priestley (1733–1804) discovered that the gas released by heating a mixture of mercury (Hg) and air also made candles burn brighter. He published his findings before Scheele. Finally, French scientist Antoine Lavoisier (1743–1794) gave oxygen its name from Greek words meaning "acid former" because many of the compounds formed by oxygen are acids. Lavoisier recognized the role of oxygen in respiration and combustion. Scheele, Priestly, and Lavoisier were part of a new generation of scientists that conducted experiments to prove their theories. They recorded their equipment, methods, and findings, much as scientists do today.

BIG THINGS ARE MADE OF LITTLE THINGS

Before we go any further, let's talk a little more about how elements fit together to create everything in our universe. For starters, each element is made up of atoms, which are the smallest pieces of an element that still have that element's properties. Each atom is of one specific element, such as an atom of carbon (C) or an atom of oxygen (O). Two or more atoms together make a molecule. Join one atom of carbon with one atom of oxygen, and you get a molecule of carbon monoxide (CO).

A DISSENTER

Joseph Priestley was educated at the Dissenting Academy in Northamptonshire, England. He was forbidden from attending an English university because of his religious beliefs. He followed a belief system known as Unitarianism. This philosophy rejected many of the more mystical elements of the Church of England and other forms of Christianity.

Unitarians believe strongly in rational thought. Priestley believed that it was important to explore the world with an open mind. Rather than blindly accepting the scientific theories of the day, he performed experiments to prove or disprove his ideas. He thought proven facts were a better basis for science than unproven theories. He designed equipment to perform his experiments. He believed that anyone was capable of discovery, not just a select few exceptionally gifted people. Priestley also believed that the problems of society could be solved by the application of logic, rational thought, and science.

Oxygen's entry on the periodic table contains lots of information. It includes its common name, chemical symbol, atomic number, and atomic weight.

Protons, neutrons, and electrons are even smaller than atoms. When oxygen forms a compound with another element, it shares the electrons in its outer shell.

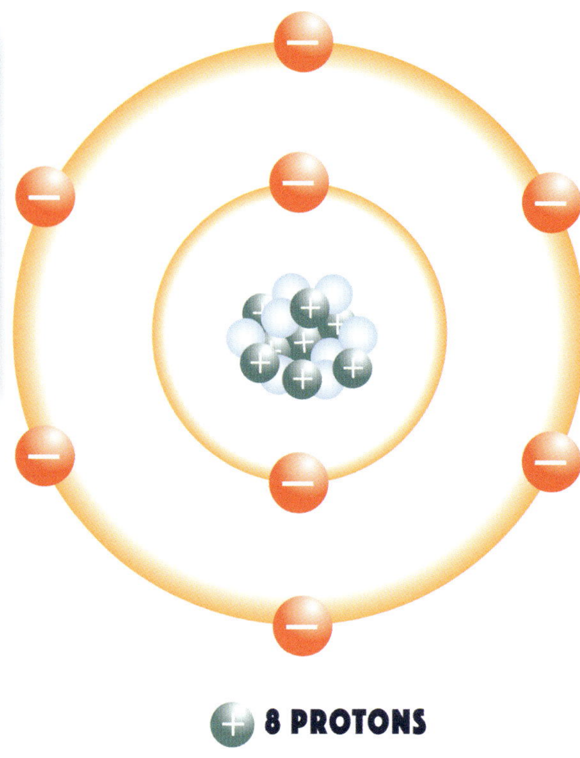

WHAT'S SMALLER THAN AN ATOM?

Atoms themselves are made up of smaller parts called subatomic particles, which include protons, electrons, and neutrons. Protons have a positive electric charge, electrons have a negative electric charge, and neutrons have no charge at all. If you were to view an atom through a very powerful microscope, you would find it is largely made up of empty space.

The nucleus, or center, of an atom contains the protons and neutrons. The nucleus is small and heavy compared to the rest of the atom. An element's atomic number usually is equal to the number of protons contained within its nucleus. Since an oxygen atom has 8 protons, its atomic number is 8. Another property of an element is its atomic weight (or atomic mass). The atomic weight is the average sum of the number of protons and neutrons in each atom of the element. Electrons are light and don't really factor in the weight. Oxygen has an atomic weight of 15.9994 (sometimes rounded up to 16).

Outside the nucleus are shells that contain the electrons. The first shell of an atom can hold only two electrons, the second shell can hold up to eight, and the third shell can hold up to 18. The oxygen atom has only two shells of electrons. The inner shell contains two electrons. The outer shell contains six electrons. The behavior of an atom is strongly influenced by the electron distribution in its shells.

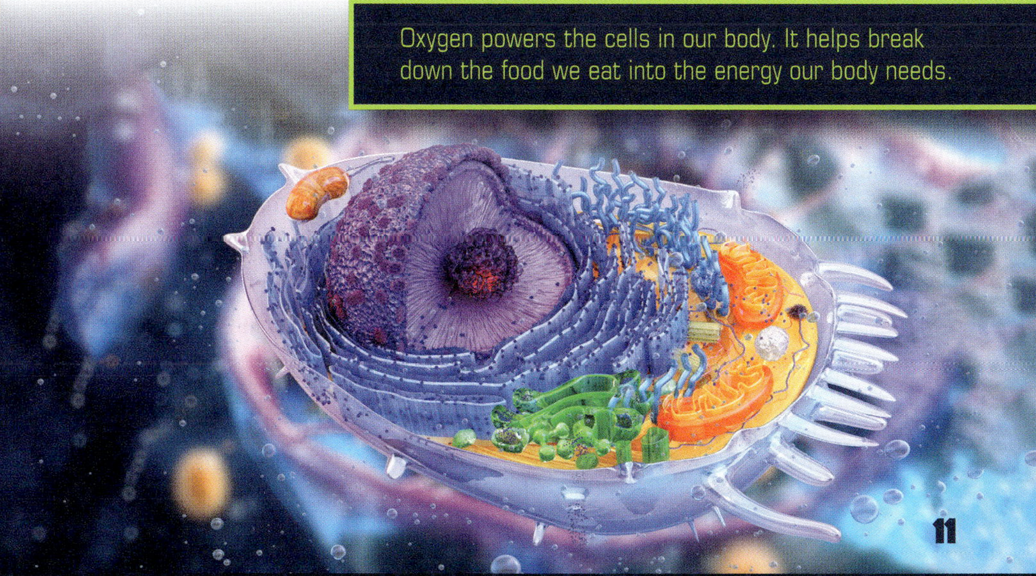

Oxygen powers the cells in our body. It helps break down the food we eat into the energy our body needs.

ISOTOPES OF OXYGEN

An atom that has the same atomic number but a different atomic weight is called an isotope. Isotopes can be stable or radioactive. Radioactive isotopes will decay or break down into other substances. For example, oxygen-17, an isotope of oxygen, contains eight protons and nine neutrons. Oxygen-18 has 10 neutrons. Oxygen-16, -17, and -18 are all stable isotopes. Radioactive isotopes of oxygen are all short-lived.

All man-made isotopes are unstable, or radioactive. They're generally created with the aid of sophisticated technology such as a particle accelerator. This includes the heaviest oxygen isotope discovered to date: oxygen-28. It contains 20 neutrons and 8 protons.

An element's atomic number is the number of protons in its nucleus. As you can see, just one proton is the difference between nitrogen and oxygen, or oxygen and fluorine.

OXYGEN

CHEMICAL SYMBOL: O

PROPERTIES: Nonmetal; odorless, colorless, tasteless

DISCOVERED BY: Independently by Joseph Priestley and Carl Scheele

ATOMIC NUMBER: 8 **ATOMIC WEIGHT:** 15.9994

PROTONS: 8 **ELECTRONS:** 8 **NEUTRONS:** 8

PHASE AT ROOM TEMPERATURE: Gas

DENSITY @ 293 K: 1.429 kg/m3

LIQUID PHASE: -297°F; -183°C; 91 K

SOLID PHASE: -360°F; -218°C; 55 K

COMMONLY FOUND: Everywhere

DIATOMIC MOLECULES

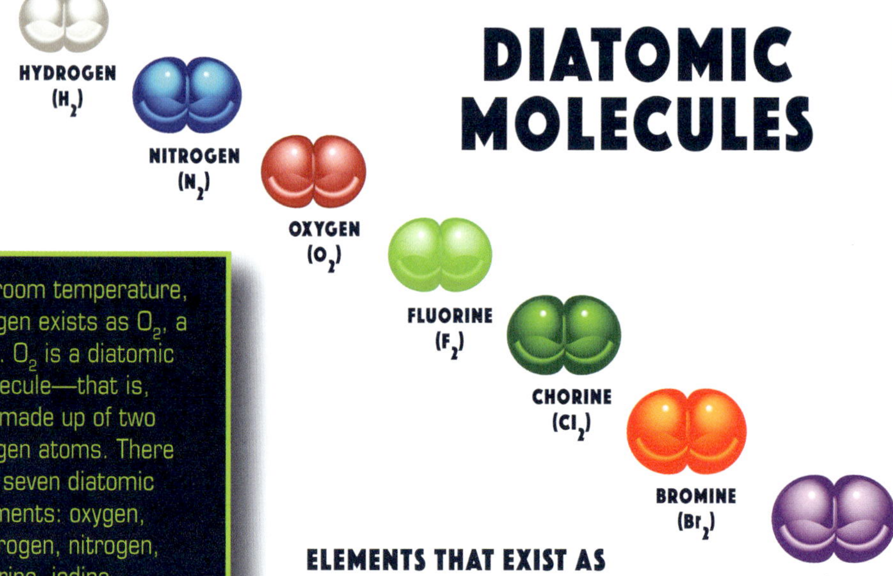

At room temperature, oxygen exists as O_2, a gas. O_2 is a diatomic molecule—that is, it's made up of two oxygen atoms. There are seven diatomic elements: oxygen, hydrogen, nitrogen, fluorine, iodine, chlorine, and bromine.

So, how does this electron distribution impact the behavior of an oxygen atom? An atom is the most stable when its electron shells are completely full. With only six electrons in its outer shell, oxygen is always on the lookout for two more electrons to complete its outer electron shell to stabilize the atom. This means that oxygen is highly reactive. In fact, most oxygen atoms in the air are combined in pairs called diatomic molecules, a name that means having two atoms. This is why oxygen almost always appears as O_2. When oxygen combines with two hydrogen (H) molecules, it forms perhaps Earth's most important compound, water (H_2O).

NO TWO ARE ALIKE

The arrangement of subatomic particles forms a sort of fingerprint for an element. No two are alike. Oxygen contains eight protons. If it was possible to remove a proton from the nucleus of an oxygen atom (it's not!) the result would be a totally different element—nitrogen (N). Adding a proton (also not possible!) would change the atom to fluorine (F). The arrangement of oxygen's electron shells is what allows it to bond easily with other elements.

Aside from the diatomic O_2 molecule, oxygen also forms an allotrope—a triatomic molecule—called O_3 or ozone. Ozone is a pale blue gas that exists in the stratosphere. It helps filter ultraviolet radiation from the sun. Located in its proper place, ozone serves a necessary function protecting life on Earth from deadly levels of radiation. However, when ozone occurs at ground level, it's a pollutant that irritates eyes and mucous membranes.

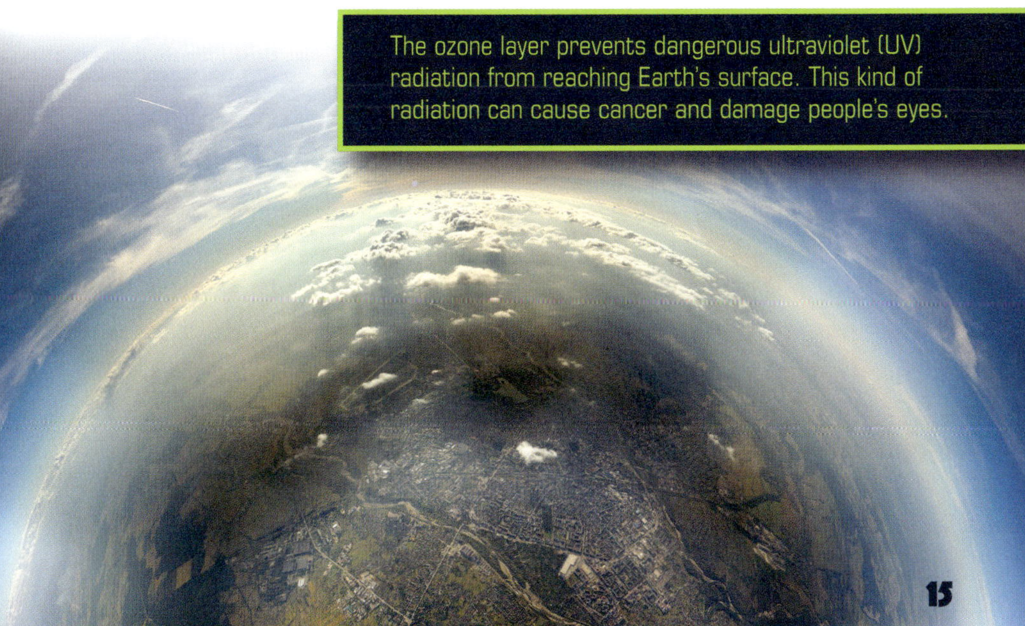

The ozone layer prevents dangerous ultraviolet (UV) radiation from reaching Earth's surface. This kind of radiation can cause cancer and damage people's eyes.

CHAPTER 2

A PLACE FOR EVERY ELEMENT

Many naturally occurring elements were identified in the 18th and 19th centuries. As people accumulated more knowledge about the natural world, they needed ways to organize all those facts. Dmitri Mendeleev (1834–1907) was a professor at the University of St. Petersburg, Russia. He'd written a textbook on organic chemistry but couldn't find one on inorganic chemistry that suited his purposes, so he began to write one too.

As he worked on his textbook, Mendeleev noticed similarities between different elements. He decided to arrange 70 known elements in a table. He organized them in horizontal rows, or periods, according to their atomic weight. The lightest element, hydrogen (H), was placed at the left and the heaviest at the right.

As he worked on his table, Mendeleev noticed gaps between known elements. Scientists of the day understood there were more elements yet to be discovered. Mendeleev believed these unknown elements would fit into the gaps of his table. His theory was proven correct during his lifetime, and his original table is the basis for the one we use today.

The modern periodic table of elements is administered by the International Union of Pure and Applied Chemistry (IUPAC). It determines the procedures for naming newly discovered elements and adding them to the table. Today, elements are placed on the periodic table according to their atomic number instead of atomic weight. Hydrogen is still number one. The elements are also organized into groups. The number of the group appears above each column of the table. By arranging the elements this way, scientists could predict where any new elements should be placed. They can also predict if an element will be a metal, a nonmetal, or a transition metal.

IUPAC approves the names of newly discovered elements before adding them to the table. This helps scientists around the world share information.

THE LATEST ADDITIONS

Mendeleev realized that there were probably more elements in the world than had been discovered. He left spaces in the periodic table for new elements to be added. His theory was proven correct, as more elements were discovered and added to the table during his lifetime. Improvements in technology have enabled the discovery or creation of additional elements. New discoveries are still being made today. The latest additions to the periodic table are four superheavy elements created in laboratories. In 2016, elements 113, 115, 117, and 118 were added to the table.

ATOMIC NUMBER	INITIAL DESIGNATION	OFFICIAL NAME
113	UNUNTRIUM (Uut)	NIHONIUM (Nh)
115	UNUNPENTIUM (Uup)	MOSCOVIUM (MC)
117	UNUNSEPTIUM (Uus)	TENNESSINE (TS)
118	UNUNOCTIUM (UUO)	OGANESSON (Og)

A replica of Mendeleev's original table is included in a monument dedicated to him in St. Petersburg, Russia.

A PLACE FOR OXYGEN

Oxygen is located within group VIA on the periodic table, which also contains sulfur (S), selenium (Se), tellurium (Te), and polonium (Po). The elements are divided by a staircase line that extends from boron (B) to astatine (At) in the lower right portion of the periodic table. This staircase line separates the metals from the nonmetals. Metals like potassium (K), sodium (Na), and iron (Fe) are located to the left of this line. Nonmetals, such as chlorine (Cl) and nitrogen, are located to the right of the line. Oxygen is located on the right side of the staircase line with the other nonmetals.

The six electrons in an oxygen atom's outer shell are what allow it to form compounds easily with so many other elements. In fact, all the other elements in the same group as oxygen on the periodic table also have six electrons in their outer shell.

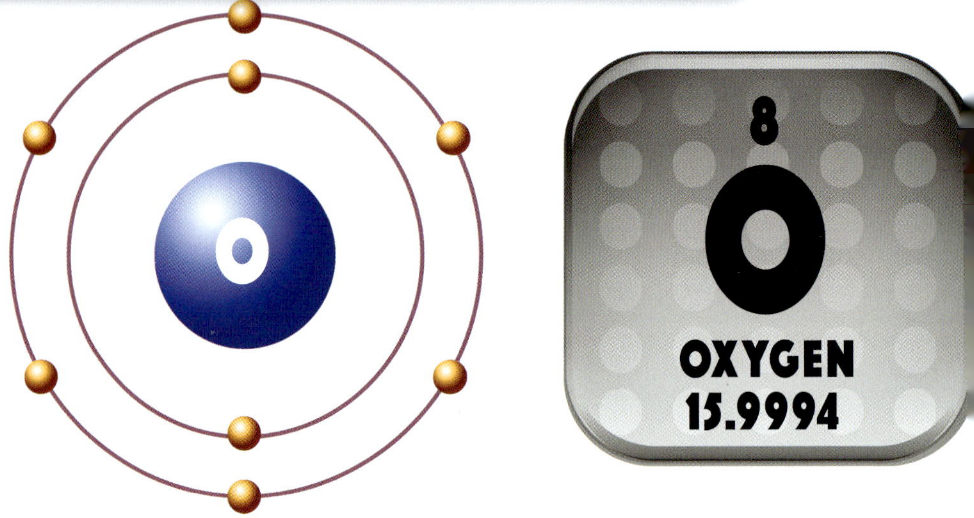

ATOMIC MASS: 15.999
ELECTRON CONFIGURATION: 2, 6

WHAT MAKES OXYGEN UNIQUE?

Every element has properties or physical characteristics that are unique to it. This includes things like its state at room temperature. All elements can appear in one of three physical states, called phases: solid, liquid, or gas. At room temperature, oxygen is a colorless, odorless gas. For oxygen to become solid, it must be cooled to a temperature of -360°F (-218°C). This is not a temperature that occurs naturally on Earth. Oxygen becomes a blueish liquid at -279°F (-183°C). This is cold enough to require special handling. The science of working with such low temperatures is called cryogenics. Liquid oxygen can be stored in pressurized containers for use for certain medical procedures or as fuel.

The air that we breathe is about 21 percent oxygen. However, sometimes doctors place patients in a hyperbaric chamber like this one so they can breathe air with a higher concentration of oxygen. This helps their body to heal from certain injuries.

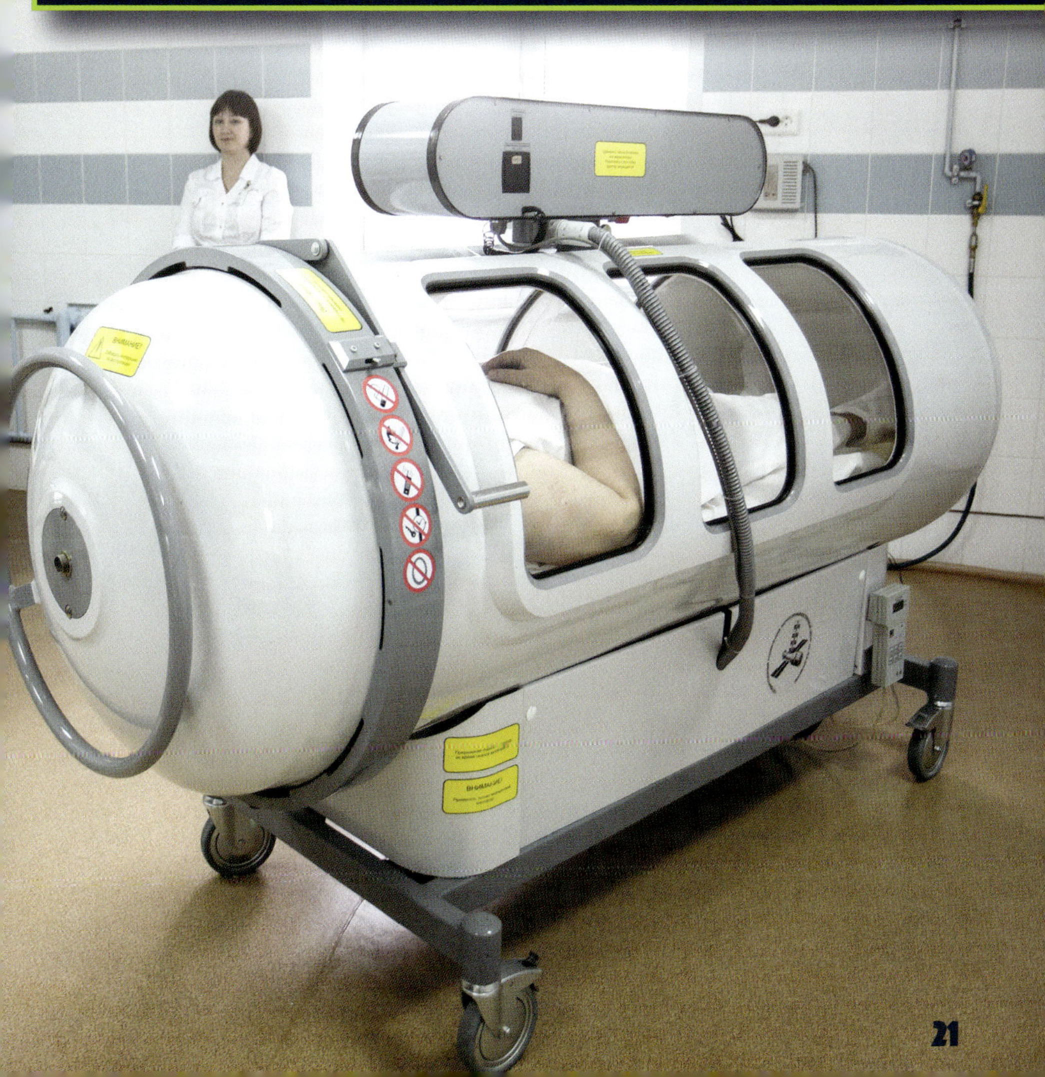

ROCKET FUEL

Oxygen is important for space travel. Not only do astronauts require oxygen to breathe, but they also need it to get to outer space. Rockets contain two liquid fuel tanks—one filled with oxygen and the other with hydrogen. Because they're both very light elements, they'd take up too much space as gases. As liquids, they're denser and therefore take up less room. Everything on a spacecraft adds to the mass that needs to be propelled out of Earth's gravity, so smaller is better.

When liquid oxygen and hydrogen combine, they ignite. The reaction between the two chemicals produces water. Because it's superheated, the water forms clouds of steam.

The oxygen astronauts breathe is stored as a gas in pressurized tanks.

CHAPTER 3

LIVING WITH OXYGEN

Everything that exists is made up of elements. Some are more important to humans than others. On Earth, oxygen is extremely important. It's the most abundant element on Earth. About 90 percent of water, 46 percent of Earth's crust, and 21 percent of its atmosphere are oxygen. Without oxygen, life as we know it could not exist. Most life-forms on Earth require oxygen to live and contain a large percentage of oxygen in their bodies. In fact, oxygen makes up two-thirds of the human body!

The oxygen-rich atmosphere that surrounds Earth protects us from dangerous levels of solar radiation. Without it, the surface of our planet might resemble Mars—empty, bare, and windy. Earth's oceans help maintain weather and temperature patterns. On Mars, where there are no oceans full of water, the temperature can vary by 90 degrees over the course of a day. The types of plant and animal life we're familiar with can't survive that kind of fluctuation. Oxygen is a big part of what makes our planet beautiful and habitable.

THE MAGIC INGREDIENT

Oxygen wasn't always so important. In fact, until just a few billion years ago, there wasn't much oxygen on this planet at all. Scientists believe that Earth started out as little more than a big rock surrounded by a cloud of gases that would be harmful to people, plants, and animal life. These gases included methane (CH_4), hydrogen, and ammonia (NH_3). A combination of substances in the rock and increasing pressure very slowly built up enough heat to melt the interior of the planet. The heavier materials, such as iron, sank toward the middle of this "earth soup," while lighter silicates, or rocks made of silicon (Si) and oxygen, rose to the surface to form Earth's earliest crust.

Does water contain more oxygen or hydrogen? Well, it depends. A water molecule contains two hydrogen atoms and one oxygen atom. However, oxygen is heavier than hydrogen, so the mass of water is mostly oxygen.

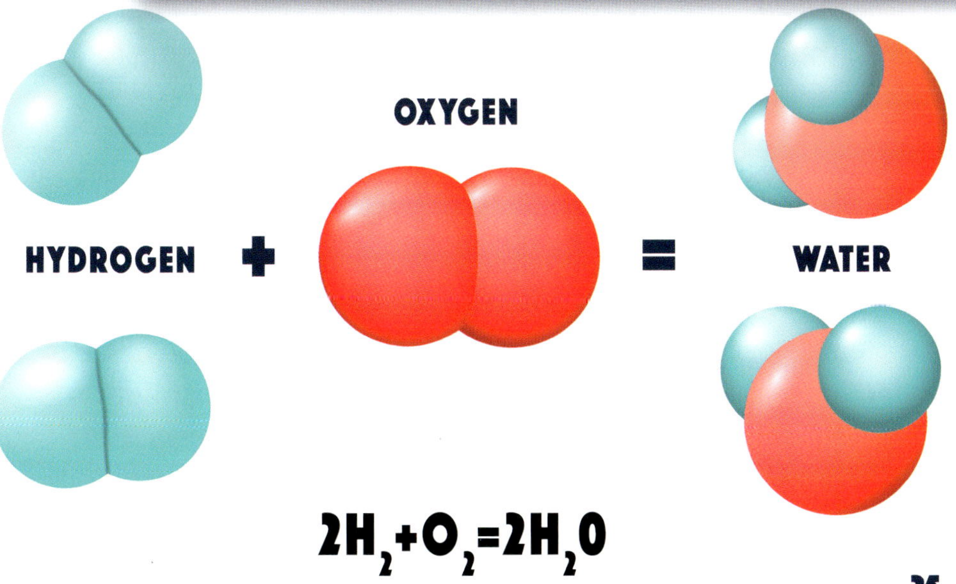

$$2H_2+O_2=2H_2O$$

Photosynthesis is a huge part of nature's recycling program. Plants take in the carbon dioxide that animals exhale and transform it into fresh supplies of oxygen.

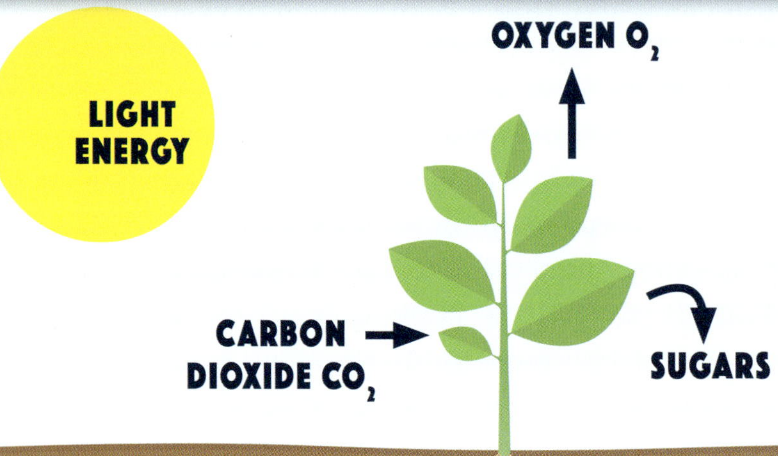

EARTH'S ATMOSPHERE

There are many theories about how Earth's atmosphere became rich with oxygen. According to one theory, the heating of Earth's interior caused other chemicals inside the planet to rise to the surface and be released into the air. Some of these chemicals formed water (H_2O), while others combined to form the gases of the atmosphere. Over millions of years, the water slowly collected in low places of the crust and formed the oceans. It is believed that oxygen first appeared in large amounts within the atmosphere nearly 3 billion years ago, when cyanobacteria, or blue-green algae, first appeared and began producing energy through the process of photosynthesis.

During photosynthesis, green plants and some bacteria use energy from the sun to combine carbon dioxide (CO_2) and water to make food. The light used in photosynthesis is absorbed by a green pigment called chlorophyll within each food-making cell. The sunlight causes water to split into molecules of hydrogen and oxygen. In a series of complicated steps, the hydrogen combines with carbon dioxide from the air, forming a simple sugar. Oxygen from the water molecules is given off in the process. It most likely took quite some time—millions of years—to build up enough oxygen in the atmosphere to support larger and more complex forms of life on Earth, including people.

Plants need CO_2 to survive. However, too much CO_2 forms pollution. Most excess CO_2 is formed due to human activity.

Earth's atmosphere acts as a sort of umbrella that protects the planet. It allows in warmth from the sun but reflects harmful levels of radiation back out into space.

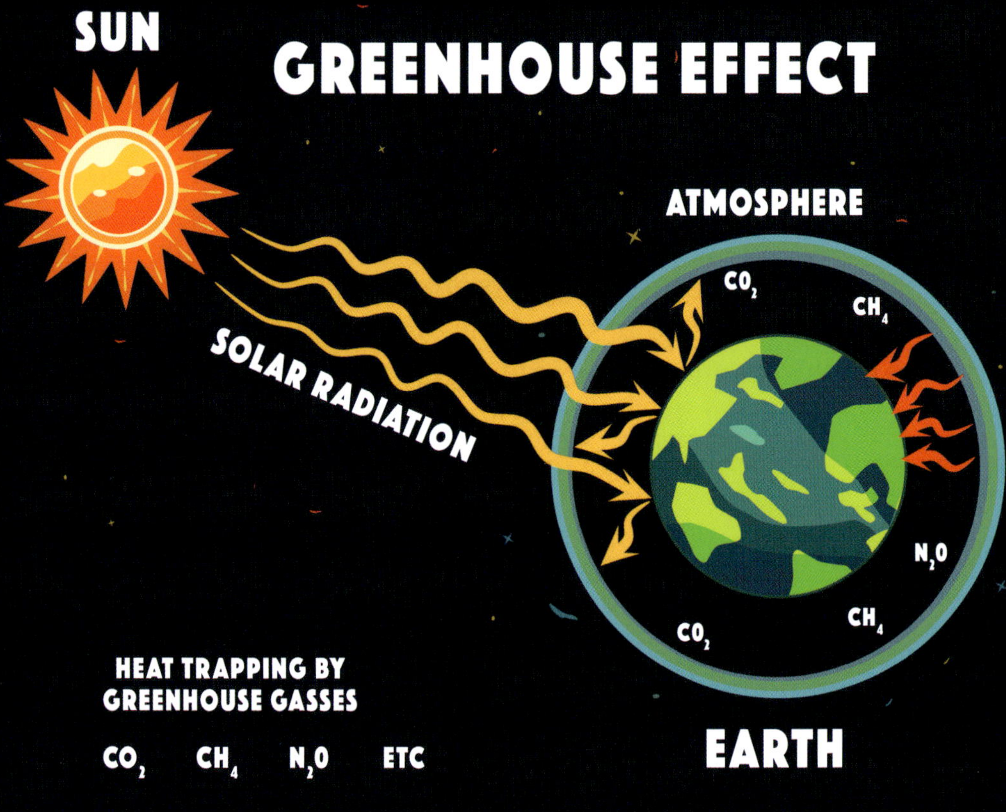

Even before people knew what oxygen was, they knew it was important to life. Throughout history, especially before the 18th century and the discovery of oxygen, many people thought that the air was made up of only one substance. As you will see, the study of oxygen is closely linked to the study of the air. In fact, it is not uncommon for people to use the two words interchangeably today, even though they are quite different.

The air is more accurately called the atmosphere, and it covers the land and sea and everything in between. It covers our whole planet and extends far above its surface. The air is invisible and has no taste or smell. Yet, it is as real as anything you can see, taste, or touch. This atmosphere—the air that we breathe—is a finely tuned mixture of gases. Oxygen makes up about 21 percent of that mixture. The remaining 79 percent of the air is a combination of several other gaseous elements.

EVERY BREATH WE TAKE

What exactly are we breathing when we inhale? Earth's atmosphere is a combination of different gases. Most of what we breathe is nitrogen (78 percent) and oxygen (21 percent). Air also contains less than 1 percent argon. But did you also know that it contains trace amounts of neon, helium, methane, krypton, hydrogen, nitrous oxide, and xenon?

The air we breathe can be damaged by pollution. This is why there are regulations about what kind of chemicals factories can release into the air. It's also why cars have to pass emissions tests.

The ozone layer is slowly healing, due to changes in human activity and regulations about what types of chemicals people can use.

OZONE

Ozone (O_3) is something you may hear mentioned in the news. But what is it, exactly? And why is it important? Ozone is a gas. It's a triatomic molecule, meaning it's made up of three oxygen atoms. Depending on where in the atmosphere it's found, ozone can be either helpful or harmful.

When ozone occurs at ground level, it's a serious form of pollution that can make people and animals sick. It irritates their eyes and mucous membranes.

Ozone exists naturally in the upper layers of the earth's atmosphere. It protects the planet—and everything that lives here—from high doses of dangerous solar radiation. However, human activities produce pollution that has thinned the ozone layer, forming holes that scientists monitor carefully. The result is that more ultraviolet radiation reaches the earth's surface. This causes sickness such as cancer and genetic mutations, as well as cataracts. In recent years, changes in chemical use have helped close many ozone holes and made others smaller.

The Montreal Protocol is an international treaty ratified in 1987. It's an agreement between 200 nations to regulate chemicals that deplete the earth's ozone layer.

CHAPTER 4

ALL ABOUT COMBUSTION

Oxygen is vitally important to life on Earth. It forms a major part of the air we breathe as well as the water we drink. It's also necessary for fire. Combustion is the exothermic chemical reaction between oxygen and other substances that results in light and heat in the form of flame.

For millennia, people have used fire for light, heat, and cooking. They've also stood in awe of its destructive capabilities. After all, the same fire that warms a house can also burn it to the ground. Ancient people experienced fire in the form of erupting volcanoes or lightning strikes that caused fires. Often, they attributed the ensuing destruction as the actions of the gods.

Even with today's technology, controlling the awesome power of fire is a daunting task. Fire is a natural part of some ecosystems. Periodic fires serve to clear out dead underbrush and provide light to lower levels of the habitat. Some plants also require fire for their seeds to germinate. However, some of these places are located where there are seasonal high winds, which can spread fire far and wide. Now that people live in places where fire occurs naturally, it can be hard to safeguard life and property.

WHAT MAKES IT BURN?

Some of the earliest efforts to understand air and fire date back many thousands of years to the philosophers and physicians of the ancient world. As early as 545 BCE, Anaximenes, a Greek philosopher, argued that the world developed out of air. He believed that air turned into other things, like water and earth, through a process called rarefaction. This process might be thought of as similar to condensation, the process by which a gas turns into a liquid. According to Anaximenes, air becomes visible through rarefaction, first as a mist or cloud, then as water, and finally as solid matter like rocks. If the air is further rarefied, he thought, it turns to fire.

Geologists have found evidence of fire (in the form of charcoal) dating back 420 million years. This is long before humans evolved.

Museums and libraries house collections that can be destroyed by water used to stop fires, so they protect their collections with fire suppression systems that cut off ventilation or replace the oxygen in a room with inert gas. Without oxygen, there can be no fire.

Around 450 BCE, Empedocles, another Greek philosopher, came to believe that air must be one of the basic elements from which all matter is created. At the time, people believed the elements were earth, fire, and water. A poet, statesman, and physician, Empedocles also believed that nothing in the universe could be created or destroyed—material could only be transformed into other objects depending on different combinations of the four essential ingredients. As we've already seen, this early scientist may have been wrong about some of the details, but he was on the right track.

One of the first people to catch on to the idea that the air is a mixture of gases is perhaps more famous for his artistic contributions than his scientific ones. "Where flame cannot live, no animal that draws breath can live," wrote Leonardo da Vinci, the legendary Italian painter and sculptor who lived during the late 15th century. He observed that a fire needs to use a part of the air around it to burn. During his study of the human body, da Vinci discovered that the same relationship plays a role in the process of respiration, or breathing. He watched the air being taken into the body with every breath as a candle burned nearby. He concluded that it was impossible for air to be an element because it was obviously composed of different parts.

Firefighters are hard-pressed to fight fires fed fresh supplies of oxygen by high winds. They use aircraft that can drop water or spray flame-retardant chemicals. On the ground, they also use construction equipment to dig firebreaks, or ditches, to try to contain the flames.

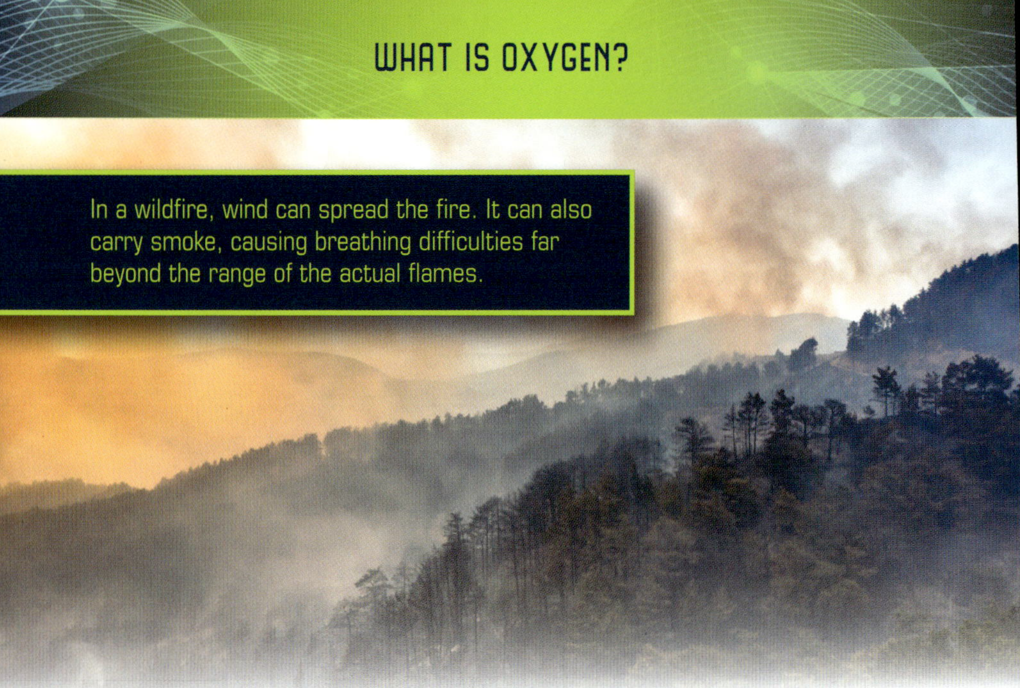

In a wildfire, wind can spread the fire. It can also carry smoke, causing breathing difficulties far beyond the range of the actual flames.

A DISPROVEN THEORY

By the start of the 18th century, scientists had different ideas about air and fire. They had figured out that the universe was made of more than water, air, fire, and earth. A German chemist and physician named Georg Ernst Stahl (1659–1734) came to believe that anything that could burn was in part composed of a substance called phlogiston. Fire, he reasoned, was caused by the release of this substance into the air. Stahl called his theory "phlogiston theory." The word "phlogiston" comes from Greek and means "burned." The ash or residue was called "dephlogisticated." The phlogiston theory was widely accepted in the 1700s and led to many findings in chemistry.

The major objection to this theory was that the ash of an organic substance, such as wood, often weighs less than the original substance. Meanwhile, calx, or metallic ash, is heavier than the metal. One of the scientists who studied these weight differences was French chemist Antoine Lavoisier. During his own experiments with fire, he realized that Scheele's "fire air" helped the calx gain weight by chemically combining with some of the air around it and not by absorbing phlogiston. Later, he noticed that this same "fire air" tended to form acids when mixed with certain substances. From the Greek words *oxys*, meaning "acid," and *genos*, meaning "forming," Lavoisier renamed the gas "oxygen." Some chemists—including Joseph Priestley, one of the discoverers of oxygen—rejected Lavoisier's findings, however, and tried to retain some form of the phlogiston theory. By 1800, practically every chemist recognized the correctness of Lavoisier's oxygen theory.

Although the phlogiston theory was disproved, it's a good example of the way science works. Scientists observed a phenomenon and proposed a theory for what was happening. Later scientists applied new technology and ideas and proposed a different theory (combustion), proven through experiments.

COMBUSTION REACTION

A SUBSTANCE BY ANY OTHER NAME

Antoine Lavoisier lived at a very interesting time in scientific history. In the 18th century, many scientists behaved more as philosophers—discussing scientific theories in the abstract. Many also still believed in the idea that everything was made of four main substances—earth, air, fire, and water.

Lavoisier proved his theories by experimentation. He discovered how water and air are formed of oxygen in combination with other elements. He also devised the system we use to name elements today. For example, oxygen is derived from Greek words meaning "acid former." Lavoisier named all the elements known to science in his time. He also arranged them into families.

Lavoisier's wife, Marie-Anne, acted as his lab assistant, taking notes and making sketches of his experiments.

In an internal combustion engine, the reaction that drives the pistons is carefully contained.

COMBUSTION IN THE MODERN WORLD

Today, we've harnessed the power of combustion to power our world. Although you may not often see an open flame, the chemical reaction of combustion occurs all the same, often contained inside an engine or generator. Combustion is the chemical reaction between a gas—usually oxygen—and a fuel source. Fuel is anything that burns, from wood or charcoal to liquid hydrogen rocket fuel.

A fire requires three things to burn: fuel, oxygen, and heat. For example, most cars operate via an internal combustion engine. In this type of engine, a spark ignites a mixture of fuel and air. The expanding gases push the piston that turns the crankshaft.

A rocket engine combines oxygen and fuel in a tank where they are ignited. The product of this kind of combustion is hot exhaust gas, which is propelled out the back of the craft to create forward thrust.

FIRE-FIGHTING EQUIPMENT

Firefighters are trained people who respond to residential and business fires and wildfires. They have equipment at their disposal to extinguish fires. The best way to put out a fire is to deprive it of one of its three necessary elements: heat, fuel, or oxygen. While you're probably most familiar with the sight of a fire engine equipped with hoses that spray water, there are other tools available depending on the type of fire.

- Fire extinguishers are designed to smother different kinds of fires. They might spray foam, powder, carbon dioxide, or special chemicals. These cool the fire, deprive it of oxygen, or both.
- Fireproof blankets can also be used to starve a fire of oxygen.
- Facilities that house valuable items such as art, books, or antiquities might employ systems that close off ventilation to an affected area or use inert chemicals to dilute the oxygen.
- Aircraft equipped with tanks of water, foam, or chemicals are used to combat forest fires.

Fuel requires a certain amount of heat to burn on its own. This is called the ignition point. Without the required level of heat, the fuel will not burn. This is why people can transport and store fuel safely. A fuel's flashpoint is the temperature at which it will ignite with a specific heat source, such as a spark.

Rockets provide a great example of how combustion works. They either use solid or liquid propellants, as shown here.

ROCKET PROPULSION

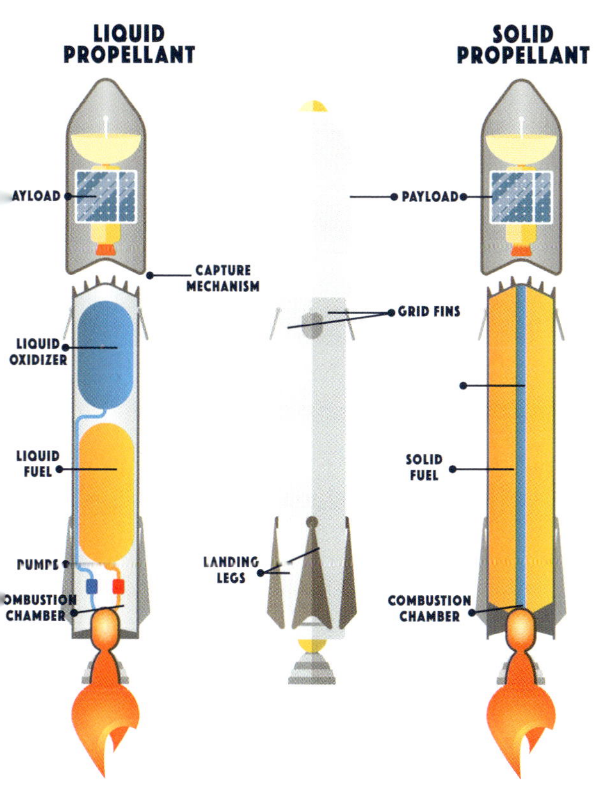

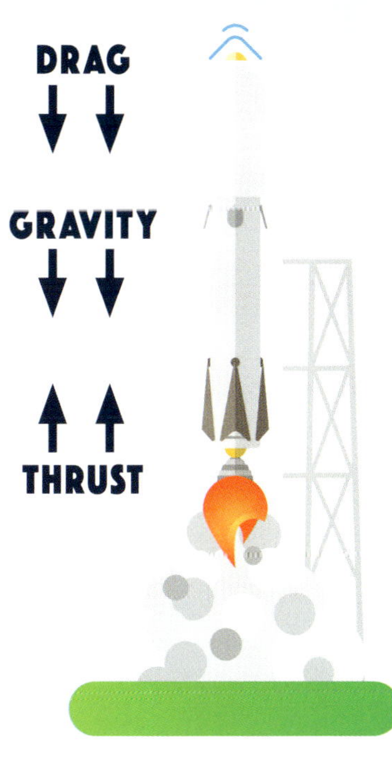

CHAPTER 5

OXYGEN MAKES FRIENDS

When different elements combine, the resulting substance is called a compound. For instance, water is the most common oxygen compound. It's a compound formed of two hydrogen atoms and one oxygen atom. Its chemical formula is H_2O. Carbon dioxide is a compound formed of one carbon and two oxygen atoms—CO_2. As you can see, a compound's chemical formula describes its composition.

Oxygen is a very reactive element, meaning it combines easily with most other elements. The exceptions are a group of elements called the noble gases. The noble gases are: helium (He), neon (Ne), argon (Ar), krypton (Kr), xenon (Xe), and radon (Rn). They're inert, meaning they don't react with other elements. For example, neon can't form compounds with other elements because its outer shell is already entirely filled with electrons.

Water isn't just the most common oxygen compound—it's the most common chemical compound on Earth.

ALL ABOUT OXIDES

Compounds of oxygen and one other element are called oxides. As Antoine Lavoisier realized, oxides are often made by heating or burning an element or a compound in the presence of air or oxygen.

Inert gases can be elemental—like the noble gases. They can also be compounds that don't readily react with other elements. Carbon dioxide and nitrogen are both inert gases.

CO_2

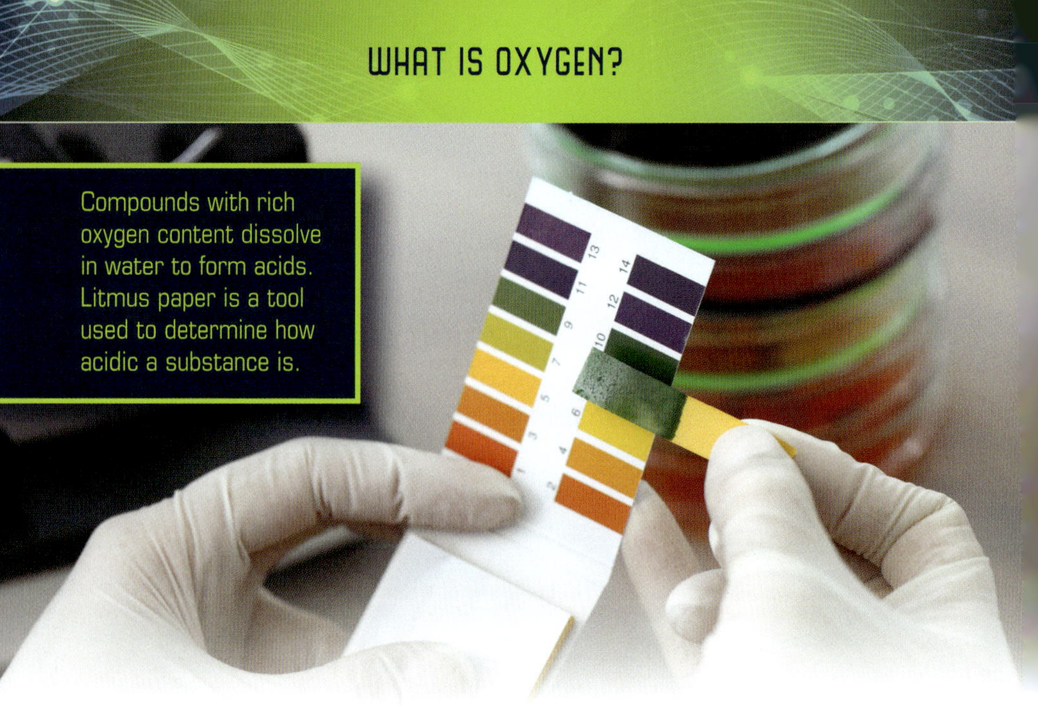

Compounds with rich oxygen content dissolve in water to form acids. Litmus paper is a tool used to determine how acidic a substance is.

OXYGEN AND METALS

When metals combine with oxygen, they form metal oxides, usually in the form of solid crystals. Magnesium oxide, for example, is white crystals, often mixed with magnesium chloride to form stucco, a light-colored cement often used to decorate the exterior of buildings.

Metal oxides are usually ionic compounds. Ionic compounds occur through the process of ionic bonding, in which the atoms of one element donate negatively charged electrons in their outer shells to the atoms of another element. In doing so, they become positively charged ions. In the case of metal oxides, the metal atoms donate electrons to oxygen atoms. The metal atom combines with an oxygen atom to form a compound. The metal atoms have become positively charged ions, while the oxygen atoms have become negatively charged ions. The opposite electrical charges cause the elements to bond together.

OXYGEN AND NONMETALS

Nonmetals combine differently with oxygen, choosing to share electrons rather than donating them to create nonmetal oxides. This process of sharing electrons is called covalent bonding, and this type of bond creates a molecule. Nonmetal oxides are usually very reactive and tend to form strong acids when dissolved in water. Their reactions are probably what Lavoisier observed when he coined the term "oxygen."

ADDING AND SUBTRACTING ELECTRONS

Oxygen is involved in two important processes: oxidation and reduction. These processes are used in many everyday activities, including burning fuels and helping animals breathe, grow, and move. Originally, scientists used the term "oxidation" to mean that a substance was gaining oxygen. The term "reduction" meant that a substance was losing oxygen. Today, scientists use these terms to describe any reaction that involves the transfer of electrons. Atoms that lose electrons are said to be oxidized, and atoms that gain electrons are said to be reduced, even when oxygen is not involved. Any substance or compound that donates oxygen or accepts electrons to cause oxidation of another compound is called an oxidizing agent. Any substance that donates electrons to cause reduction is called a reducing agent.

Rust is probably the most familiar form of oxidation. Oxidation occurs when a molecule or atom loses electrons. Rust is the result of a chemical reaction between iron, water, and oxygen.

GETTING RUSTY

Rusting, for example, is an oxidation process. When exposed to air, most metals oxidize, forming a dull metal oxide coating or tarnish. In the case of aluminum (Al) or silver (Ag), this coating protects the untarnished metal below it from the air and further corrosion. However, when iron and steel are exposed to oxygen, a layer of brown iron oxide forms on the surface of the metal. This layer does not protect the metal below it from air and moisture. The metal merely crumbles away, and the rust can quickly "eat" through the metal. The only way to protect against this process is by applying a protective coating that seals the metal from air and moisture.

OXIDATION AND... LAW ENFORCEMENT?

A breathalyzer is an important tool used by law enforcement professionals. It's a portable device they use to test the blood alcohol level of drivers involved in accidents or moving violations. Police administer a breathalyzer test at the scene to determine immediately if the driver of the vehicle has consumed too much alcohol to drive safely.

A breathalyzer works by analyzing the air a person exhales. Police will direct a person to exhale into the tube of the breathalyzer. A fuel cell in the breathalyzer oxidizes the alcohol on their breath. This produces an electrical current. The higher the concentration of alcohol on their breath, the greater the current.

Blood alcohol content can also be determined by a blood or urine test. However, these cannot be administered at the scene of an accident or suspected driving infraction. A breathalyzer test allows law enforcement personnel to record the level of alcohol in a person's system very close to the time of an accident or observed unsafe driving.

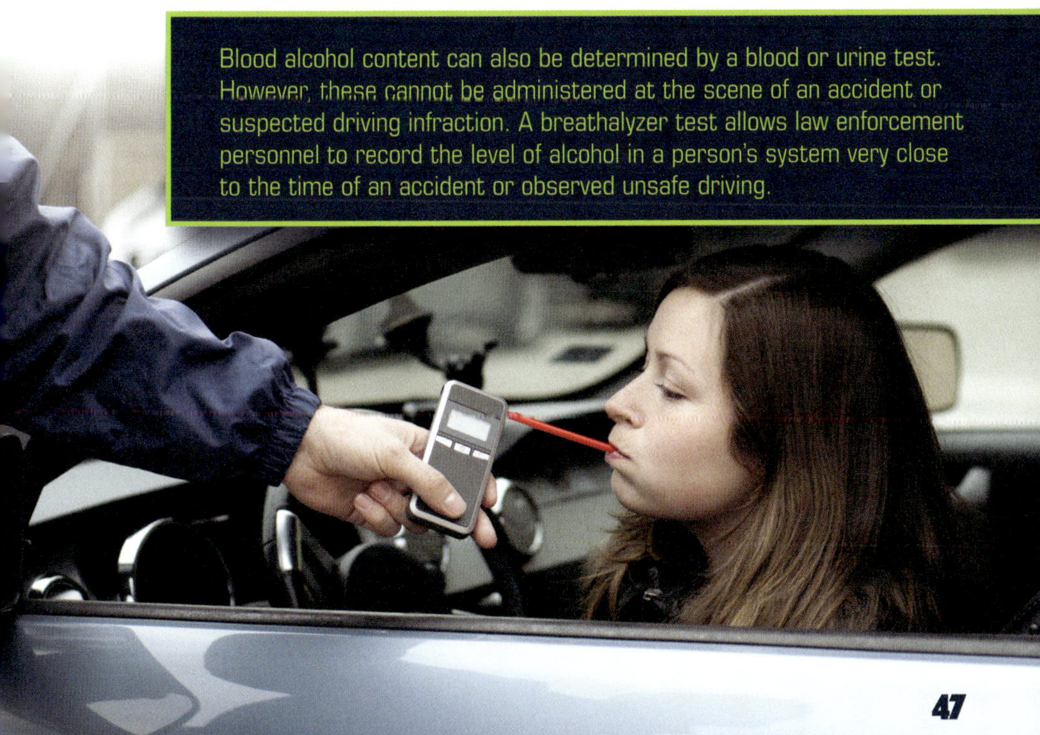

CHAPTER 6

OXYGEN EVERY DAY

We know that oxygen is vital to human and animal life, but what exactly does it do? Our bodies take in oxygen when we inhale. Our lungs are lined with tiny blood vessels that transfer oxygen from the air we breathe into red blood cells. Red blood cells contain an iron-based protein called hemoglobin that helps move oxygen through the bloodstream, ensuring it gets to all the tissues that use it to carry out an important chemical process called cellular respiration.

Oxygen helps extract energy from the food you eat. The process of digestion turns the nutrients in the food you eat into a sugar called glucose. For cellular respiration to occur, your body requires fuel (the glucose extracted from the food you ate) and an electron acceptor. Oxygen is the most efficient electron acceptor. It helps your cells carry out a form of combustion. The glucose in your cells is burned, creating energy your body needs to function. This includes the energy your body needs to carry out the process of digestion to get more glucose! Cellular respiration releases water, which your body uses. It also releases carbon dioxide, which is exhaled as a waste product.

The cardiovascular system transports oxygen all around your body. Arteries carry oxygen-rich blood from your heart to your whole body. Veins carry carbon dioxide back to your lungs to be expelled when you breathe out.

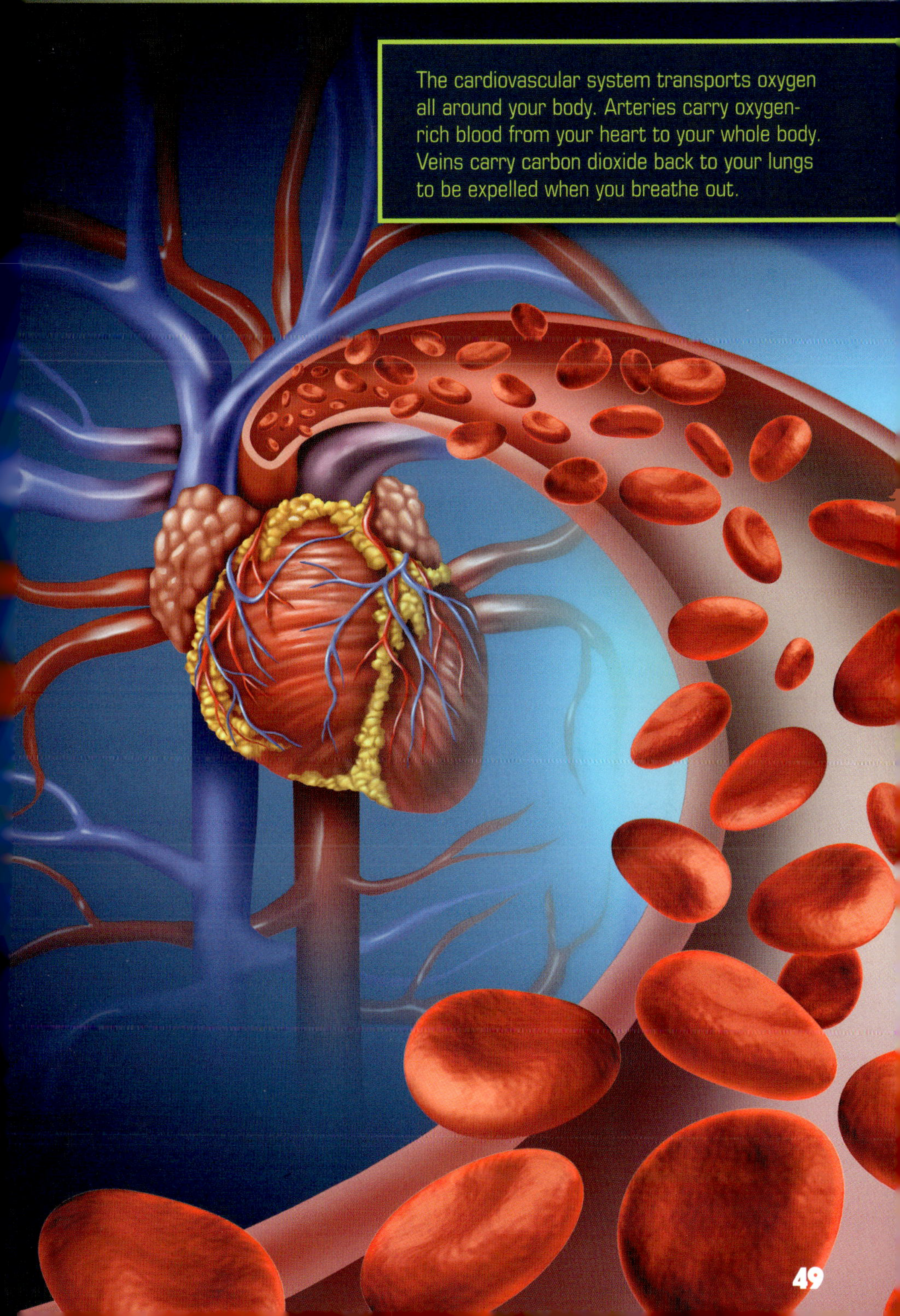

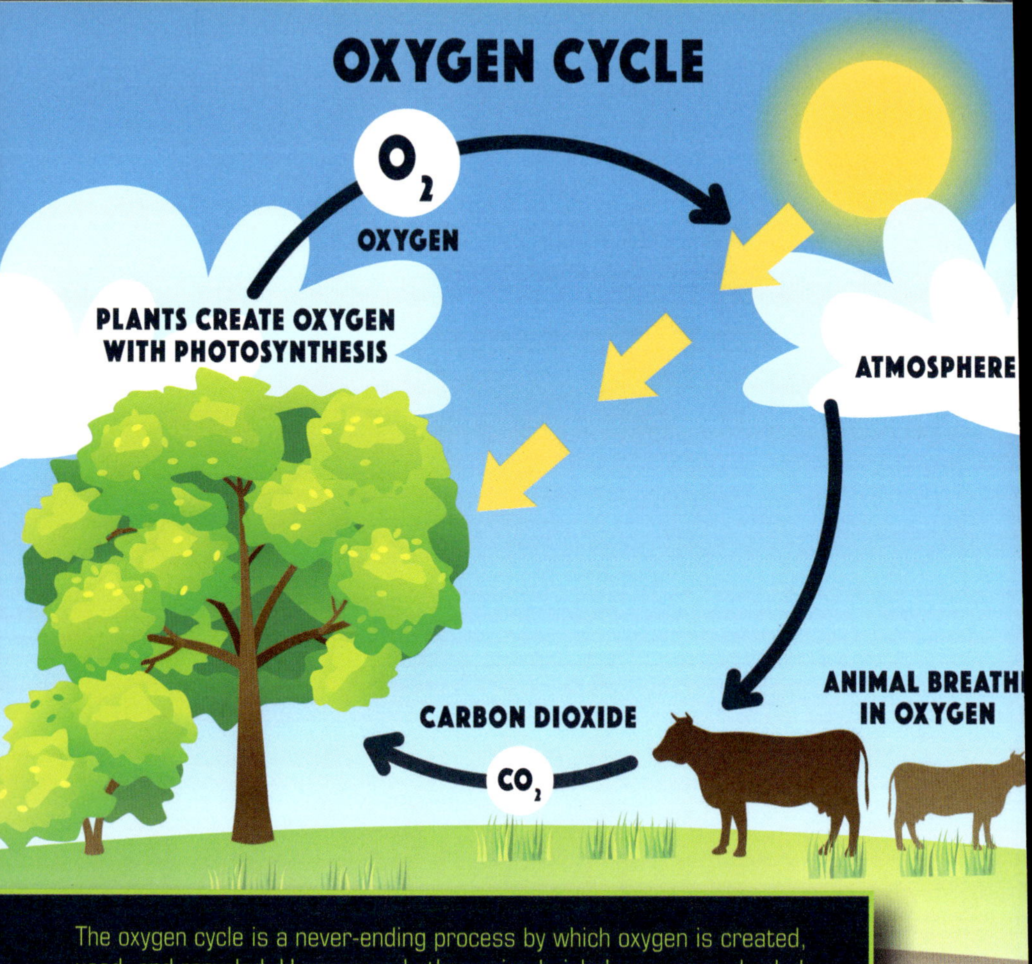

The oxygen cycle is a never-ending process by which oxygen is created, used, and recycled. Humans and other animals inhale oxygen and exhale carbon dioxide. Plants use carbon dioxide, sunlight, and water to create the energy they need. Then they release oxygen back into the air.

A NATURAL CYCLE

Inhaling oxygen and exhaling carbon dioxide are part of a larger process called the oxygen cycle. The oxygen cycle involves a constant exchange of air and oxygen between animals and plants. Animals, including humans, inhale oxygen from the air and exhale carbon dioxide. Plants absorb the carbon dioxide and release oxygen into the air through photosynthesis. Photosynthesis is the process by which green plants create food. Without plants, we wouldn't be able to breathe. Without animals breathing, plants wouldn't be able to produce oxygen. And without oxygen, there would be no life on Earth.

PORTABLE OXYGEN

After oxygen's discovery, one of its earliest uses was in medicine. Hospitals still use bottled oxygen to help patients whose bodies are not strong enough to take it from the air. Patients may be given pure oxygen, which goes right into their bloodstreams and helps them live. Firefighters also use bottled oxygen in situations where the air is too dangerous to breathe normally. By carrying an oxygen tank and gas mask, firefighters can breathe inside a fire

Most bottled oxygen is used for medical purposes.

scene. Here, most of the oxygen is being used to fuel the fire. There is little oxygen left over, so breathing can be very difficult. Bottled oxygen is also used by climbers who travel to the highest mountain peaks, where the air is thin and oxygen is scarce. Divers also carry oxygen tanks to help them stay underwater for long periods of time.

INDUSTRIAL OXYGEN

Oxygen is also used in heavy industrial processes, such as manufacturing steel. Pure oxygen is injected at very high speeds into a blast furnace containing liquid iron and scrap metal. The oxygen quickly oxidizes the elements in the furnace. Oxygen is also used for cutting and welding molten metals.

Oxygen intended for industrial use isn't the same as medical oxygen. Industrial oxygen is used to speed up chemical reactions. It's not regulated by the Food and Drug Administration (FDA) like medical oxygen, and it's not safe to breathe.

As you can see, oxygen is all around us. It's in the air and water that sustain us and make Earth beautiful. It's in hospitals helping save lives and inside our bodies fueling life processes. Oxygen powers the natural and industrial world.

Humans are an important part of the oxygen cycle. Plants are important, too, which is why protecting trees is good for the air we breathe.

THE PERIODIC TABLE OF ELEMENTS

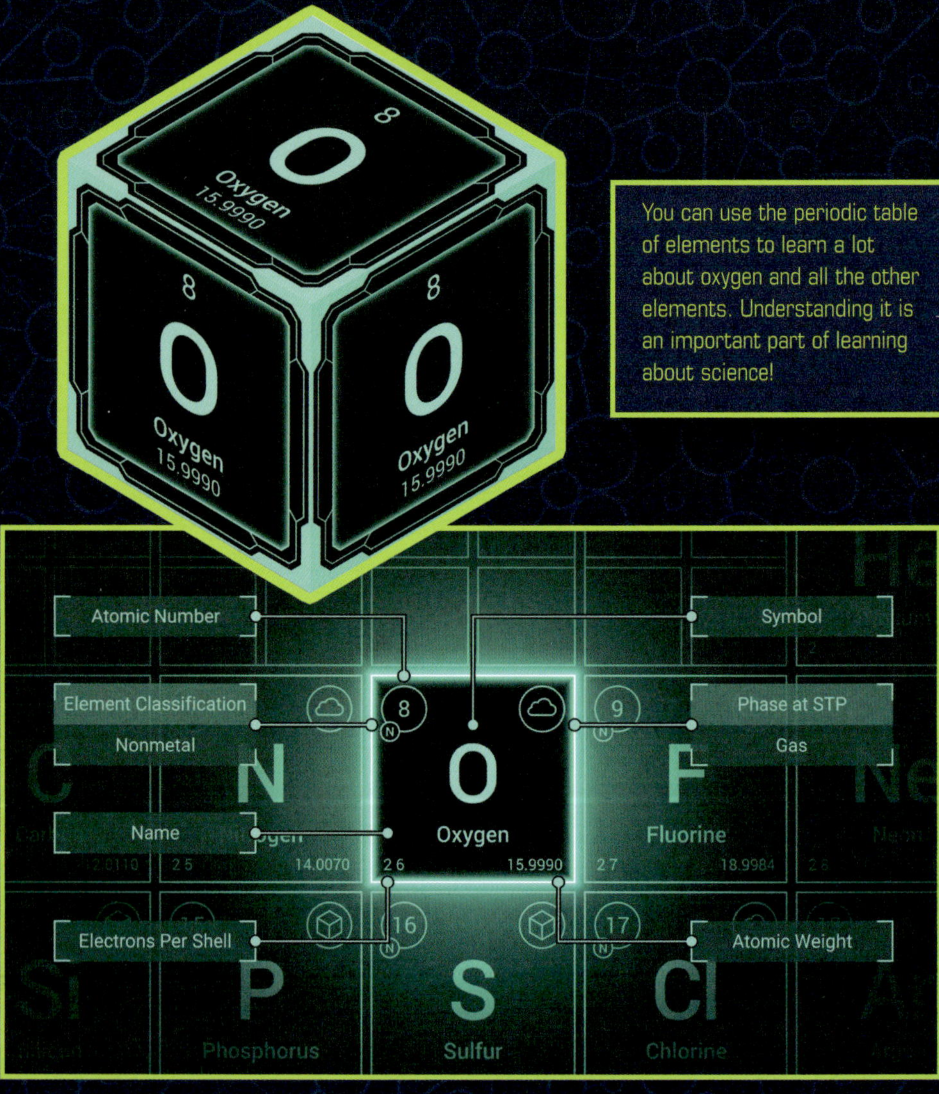

You can use the periodic table of elements to learn a lot about oxygen and all the other elements. Understanding it is an important part of learning about science!

PERIODIC TABLE OF ELEMENTS

- Alkali metals
- Alkaline earth metals
- Transition metals
- Metal
- Metalloids
- Nonmetals
- Halogen
- Noble gases
- Lanthanoids
- Actinoids

1	2	3	4	5	6	7	8	9	10	11	12	13	14	15	16	17	18
1 H Hydrogen 1.008																	2 He Helium 4.0026
3 Li Lithium 6.94	4 Be Beryllium 9.0122											5 B Boron 10.81	6 C Carbon 12.011	7 N Nitrogen 14.007	8 O Oxygen 15.999	9 F Fluorine 18.998	10 Ne Neon 20.180
11 Na Sodium 22.990	12 Mg Magnesium 24.305											13 Al Aluminium 26.982	14 Si Silicon 28.085	15 P Phosphorus 30.974	16 S Sulfur 32.06	17 Cl Chlorine 35.45	18 Ar Argon 39.948
19 K Potassium 39.098	20 Ca Calcium 40.078	21 Sc Scandium 44.956	22 Ti Titanium 47.867	23 V Vanadium 50.942	24 Cr Chromium 51.996	25 Mn Manganese 54.938	26 Fe Iron 55.845	27 Co Cobalt 58.933	28 Ni Nickel 58.693	29 Cu Copper 63.546	30 Zn Zinc 65.38	31 Ga Gallium 69.723	32 Ge Germanium 72.630	33 As Arsenic 74.922	34 Se Selenium 78.971	35 Br Bromine 79.904	36 Kr Krypton 83.798
37 Rb Rubidium 85.468	38 Sr Strontium 87.62	39 Y Yttrium 88.906	40 Zr Zirconium 91.224	41 Nb Niobium 92.906	42 Mo Molybdenum 95.95	43 Tc Technetium (98)	44 Ru Ruthenium 101.07	45 Rh Rhodium 102.91	46 Pd Palladium 106.42	47 Ag Silver 107.87	48 Cd Cadmium 112.41	49 In Indium 114.82	50 Sn Tin 118.71	51 Sb Antimony 121.76	52 Te Tellurium 127.60	53 I Iodine 126.90	54 Xe Xenon 131.29
55 Cs Caesium 132.91	56 Ba Barium 137.33	57-71	72 Hf Hafnium 178.49	73 Ta Tantalum 180.95	74 W Tungsten 183.84	75 Re Rhenium 186.21	76 Os Osmium 190.23	77 Ir Iridium 192.22	78 Pt Platinum 195.08	79 Au Gold 196.97	80 Hg Mercury 200.59	81 Tl Thallium 204.38	82 Pb Lead 207.2	83 Bi Bismuth 208.98	84 Po Polonium (209)	85 At Astatine (210)	86 Rn Radon (222)
87 Fr Francium (223)	88 Ra Radium (226)	89-103	104 Rf Rutherfordium (267)	105 Db Dubnium (268)	106 Sg Seaborgium (269)	107 Bh Bohrium (270)	108 Hs Hassium (277)	109 Mt Meitnerium (278)	110 Ds Darmstadtium (281)	111 Rg Roentgenium (282)	112 Cn Copernicium (285)	113 Nh Nihonium (286)	114 Fl Flerovium (289)	115 Mc Moscovium (290)	116 Lv Livermorium (293)	117 Ts Tennessine (294)	118 Og Oganesson (294)

57 La Lanthanum 138.91	58 Ce Cerium 140.12	59 Pr Praseodymium 140.91	60 Nd Neodymium 144.24	61 Pm Promethium (145)	62 Sm Samarium 150.36	63 Eu Europium 151.96	64 Gd Gadolinium 157.25	65 Tb Terbium 158.93	66 Dy Dysprosium 162.50	67 Ho Holmium 164.93	68 Er Erbium 167.26	69 Tm Thulium 168.93	70 Yb Ytterbium 173.05	71 Lu Lutetium 174.97
89 Ac Actinium (227)	90 Th Thorium 232.04	91 Pa Protactinium 231.04	92 U Uranium 238.03	93 Np Neptunium (237)	94 Pu Plutonium (244)	95 Am Americium (243)	96 Cm Curium (247)	97 Bk Berkelium (247)	98 Cf Californium (251)	99 Es Einsteinium (252)	100 Fm Fermium (257)	101 Md Mendelevium (258)	102 No Nobelium (259)	103 Lr Lawrencium (266)

- C Solid
- Hg Liquid
- H Gas
- Rf Unknown

GLOSSARY

allotrope: A substance and especially an element in two or more different forms (as of crystals) usually in the same phase.
atmosphere: The layer of gases around an object in space. On Earth, this layer is air.
atomic number: Generally, the number of protons that exist within the nucleus of an atom. The atomic number determines an element's place on the periodic table.
combustion: The process of combining with oxygen to burn.
cryogenics: A branch of physics that deals with the production and effects of very low temperatures.
diatomic: Consisting of two atoms; having two atoms in the molecule.
electron: A negatively charged particle found within an atom but outside of the nucleus.
exothermic: Characterized by or formed with evolution of heat.
ion: A charged atom.
isotope: An atom of a chemical element with the same atomic number and nearly identical chemical behavior but with a different atomic weight and different physical properties.
mass: The measure of the amount of matter in something.
matter: What things are made of. Matter takes up space and has mass.
molecule: Two or more atoms bonded together.

mucous membrane: A lining of body cavities and passages (as of the gastrointestinal or respiratory tract) that produce a thick, slippery liquid to protect the body.

neutron: An atomic particle that has no electrical charge and is found inside the nucleus.

nucleus: The center of an atom that contains protons and neutrons.

oxidation: The process by which oxygen is added to something or by which atoms donate electrons.

photosynthesis: The process in which green plants make their own food from sunlight, water, and carbon dioxide.

proton: A subatomic particle that has a positive electrical charge and is found inside the nucleus.

reduction: The process by which oxygen is removed from something or by which atoms gain electrons.

triatomic: Having three atoms in the molecule.

volume: The amount of space that something occupies.

FOR MORE INFORMATION

AMERICAN CHEMICAL SOCIETY (ACS)

1155 16th Street NW
Washington, DC 20036
(800) 227-5558
Website: www.acs.org
Facebook: @AmericanChemicalSociety
Instagram and X: @AmerChemSociety
ACS supports STEM education through grants and scholarships. ACS members represent 140 countries.

CHEMICAL INSTITUTE OF CANADA (CIC)

90-2420 Bank Street
Ottawa, Ontario, Canada
K1V 8S1
(613) 232-6252
Website: www.cheminst.ca/
X: @CIC_ChemInst
The CIC facilitates the research and cooperation of chemists and chemical engineers in Canada.

INTERNATIONAL UNION OF PURE AND APPLIED CHEMISTRY (IUPAC)

IUPAC Secretariat
79 T.W. Alexander Drive
Research Commons Building 4501, Suite 190
Research Triangle Park, NC 27709, USA
(919) 485 8700
Website: iupac.org/
Facebook: @iupac.org
X: @iupac
IUPAC is the global chemistry organization that maintains the periodic table of elements.

INTERNATIONAL UNION OF PURE AND APPLIED PHYSICS (IUPAP)

Fondazione Internazionale Trieste per il Progresso e la Libertà delle Scienze
c/o ICTP
Strada Costiera 11
34151 Trieste
Website: iupap.org/
Facebook: @iupap
X: @iupap_physics
IUPAP's mission is to assist in the global advancement of physics and facilitate international scientific cooperation.

NATIONAL AERONAUTICS AND SPACE ADMINISTRATION (NASA)

Mary W. Jackson NASA Headquarters
300 E. Street SW, Suite 5R30
Washington, DC 20546
(202) 358-0001
Website: www.nasa.gov
Facebook, Instagram, and X: @NASA
NASA is the U.S. government agency responsible for air and space exploration.

NATIONAL GEOGRAPHIC SOCIETY

1145 17th Street NW
Washington, DC 20036
(202) 857-7700
Website: www.nationalgeographic.com
Facebook, Instagram, and X: @natgeo
This is one of the most prestigious organizations for exploration and education about history, human society, and the natural world.

FOR FURTHER READING

Chanquet, Alain. *Oxygen: The Chemical Adventure of Léo and Zoé*. Independently published, 2024.

Chiran, Carmen, and David Chiran. *Oxygen's Vital Quest: Keeping You Alive!: Oxy's Incredible Story of Transformation From Gas to Energy in The Human Body*. Independently published, 2024.

Congdon, Lisa. *The Illustrated Encyclopedia of the Elements: The Powers, Uses, and Histories of Every Atom in the Universe.* San Francisco, CA: Chronicle Books, LLC, 2021.

McHenry, Ellen Johnston. *The Chemical Elements Coloring & Activity Book: A Fun and Interactive Guide to the World of Chemistry*. Independently Published, 2021.

Emminizer, Theresa. *The Oxygen Cycle.* Buffalo, NY: Enslow Publishing, 2023.

FOR FURTHER READING

Suarez, Nat. *Oxygen: The Lonely Element.* Independently published, 2022.

Thomas, Isabel, and Sara Gillingham. *Exploring the Elements: A Complete Guide to the Periodic Table.* New York, NY: Phaidon Press, Inc. 2021.

Zovinka, Edward P., and Rose A. Clark. *A Kids' Guide to the Periodic Table: Everything You Need to Know About the Elements.* Emeryville, CA: Rockridge Press, 2020.

INDEX

ABOUT THE AUTHOR

KATHLEEN A. KLATTE is the author of many nonfiction books for children and teens. Topics she has written about range from animals and nature to constitutional law to unusual career choices. She lives in New York with one cat and far too many books and Legos.

PHOTO CREDITS

Cover, p. 1 remotevfx.com/Shutterstock.com; Cover, pp. 1, 3-64 Oksancia/Shutterstock.com; Cover, pp. 1, 3-64 pluie_r/Shutterstock.com; p. 5 Sergey Fedoskin/Shutterstock.com; p. 7 PeopleImages.com - Yuri A/Shutterstock.com; p. 9 Srikanth_K/Shutterstock.com; p. 10 tersetki/Shutterstock.com; p. 11 Corona Borealis Studio/Shutterstock.com; p. 12 Ralf Juergen Kraft/Shutterstock.com; p. 13 Peter Hermes Furian/Shutterstock.com; p. 14 OSweetNature/Shutterstock.com; p. 15 Bilanol/Shutterstock.com; p. 17 bangoland/Shutterstock.com; p. 19 Srg Gushchin/Shutterstock.com; p. 20 BlueRingMedia/Shutterstock.com; p. 21 Slutsky Maksim/Shutterstock.com; p. 22 Maria Made/Shutterstock.com; p. 23 Blueee77/Shutterstock.com; p. 25 Designua/Shutterstock.com; p. 26 Misha Mishchenko/Shutterstock.com; p. 27 LookerStudio/Shutterstock.com; p. 28 Ozant/Shutterstock.com; p. 29 Antonio Guillem/Shutterstock.com; p. 30 PunyaFamily/Shutterstock.com; p. 31 Artsiom P/Shutterstock.com; p. 33 Ashish Manohar Tarar/Shutterstock.com; p. 34 Alex Segre/Shutterstock.com; p. 35 Toa55/Shutterstock.com; p. 36 ekrem sahin/Shutterstock.com; p. 37 Anshuman Rath/Shutterstock.com; p. 38 https://commons.wikimedia.org/wiki/File:Lavoisier_humanexp.jpg; p. 39 VFXartist/Shutterstock.com; p. 40 New Africa/Shutterstock.com; p. 41 VectorMine/Shutterstock.com; p. 42 aNaoki Kim/Shutterstock.com; p. 43 ProStockStudio/Shutterstock.com; p. 44 H_Ko/Shutterstock.com; p. 46 tomeqs/Shutterstock.com; p. 47 nikamo/Shutterstock.com; p. 49 Lightspring/Shutterstock.com; p. 50 VectorMine/Shutterstock.com; p. 51 AlteredR/Shutterstock.com; p. 53 XiXinXing/Shutterstock.com.